MULTIPLE MINI INTERVIEW (MMI) PRACTICE QUESTIONS

500+ Practice Interview Questions for pre-healthcare including Medicine, Pharmacy, Dentistry, Physiotherapy, Midwifery, Optometry, Nursing, Veterinary Medicine, Osteopathic Medicine, Dietetics, Speech and Language Pathology, Occupational Therapy, Paramedicine, Respiratory Therapy, and Physician Assistants

JACQUELYN CRAGG

Multiple Mini Interview (MMI) Practice Questions
Copyright © 2023 by Jacquelyn Cragg

All rights reserved. No part of this publication may be reproduced, distributed, or transmitted in any form or by any means, including photocopying, recording, or other electronic or mechanical methods, without the prior written permission of the author, except in the case of brief quotations embodied in critical reviews and certain other non-commercial uses permitted by copyright law.

Tellwell Talent
www.tellwell.ca

ISBN
978-1-998190-13-3 (Hardcover)
978-1-998190-12-6 (Paperback)
978-1-998190-14-0 (eBook)

Table of Contents

Introduction 1

Chapter 1: Healthcare Professional Schools Topics 3

Chapter 2: Biomedical Topics 9

Chapter 3: Pharmaceutical Topics 35

Chapter 4: Academic/Research Topics 37

Chapter 5: Equity, Diversity, and Inclusion (EDI) and Cultural Sensitivity Topics 43

Chapter 6: General University Topics 47

Chapter 7: Media & Technology Topics 51

Chapter 8: General Topics 57

Chapter 9: Interpretation & Teaching Topics 77

About the Author 81

Introduction

The multiple mini interview (MMI) is a structured interview that has been used by a growing number of universities, colleges, and other institutions as part of their admissions process in recent years. The MMI is used in a variety of fields, including medicine, pharmacy, nursing, physiotherapy, dentistry, veterinary medicine, and other health professions, as well as business and law. It has also been used as a tool for selecting candidates for employment in a variety of settings.

The MMI consists of a series of short, standardized interviews, each lasting about seven or eight minutes. These interviews are typically conducted in a circuit format, with candidates moving from one station to the next. At each station, the candidate is given a prompt or question. The interviewer then evaluates the candidate's response based on a set of predetermined criteria. MMI assessment criteria typically include communication skills, problem-solving abilities, ethical decision-making, empathy, and professionalism. The MMI can be delivered in a virtual format or in-person format.

One of the key benefits of the MMI is that it allows for a more consistent and fair evaluation of candidates, as each station is structured with the same prompt and follow-questions and evaluated in the same way. Also, MMI evaluators are not aware of a candidate's responses to previous questions, so the candidate has a fresh start at each station. This contrasts with traditional one-on-one interviews or panel interviews, which can be less structured and less forgiving. Another benefit of the MMI is that it allows for a more diverse assessment of the candidate, since typically around 10 individuals (e.g., established healthcare professionals, healthcare students, community members, academics) from a variety of settings make up the MMI interview team. One of the challenges of the MMI is that it can be time-consuming and resource-intensive to implement, as it requires the development of a large number of standardized prompts/questions, and the training of a team of evaluators to score the responses of candidates. It also requires candidates to be able to adapt to a variety of different scenarios and topics in a short period of time, which can be challenging for some individuals. Despite these challenges, the MMI has gained widespread acceptance as a valuable tool for evaluating candidates. It has been shown to be a reliable and valid predictor of academic and professional success and has been endorsed by several professional organizations.

Below you will find hundreds of practice questions formatted in the following way: prompt, main prompt question in **bold**, and follow-up questions in *italics*. Some of the practice questions are related to health, health care, and health systems. Since MMIs are designed to assess communication and critical thinking more than subject knowledge, many of the practice questions are not related to health; they cover a wide range of topics from technology to current events. We recommend you use the practice questions to simulate a typical MMI, where you have two to three minutes to read the first question prompt (in **bold**), and recommend having a peer or family member read the follow-up questions (in *italics*) to you after you have answered the main prompt, since the follow-up questions are typically not available in written form to you. Sometimes you might answer some of the follow-up questions in your initial response; that is great if you are able to anticipate some of the follow-up questions.

Best of luck on your journey towards becoming a healthcare professional!

Chapter 1: Healthcare Professional Schools Topics

1. For most healthcare professional schools, volunteering is considered an important part of non-academic qualities in a candidate. While volunteering is generally seen as a positive, since volunteering can have an altruistic component (reflecting a true concern for the welfare of others), there is also a potential egoistic component, in that the volunteer receives clear benefits.

Do you think volunteering has altruistic components?

How do you think volunteering has egoistic components? Should volunteering be a requirement for healthcare professional schools? How can volunteering help to address social and community issues? What are effective strategies for ensuring that volunteer efforts are impactful and sustainable? How does volunteering impact academic performance, and what are some effective strategies for balancing volunteer commitments with academic demands? How does volunteering intersect with issues of privilege and power, particularly for students who may have greater access to resources and opportunities?

2. Shadowing healthcare professionals can offer valuable insights to help determine whether a career in a certain area of healthcare is a good fit.

What are other potential benefits of job shadowing in healthcare fields?

What are potential drawbacks of job shadowing in healthcare fields? What are other ways students can determine whether healthcare is a good fit? Should healthcare professional schools have formal policies regarding job shadowing?

---

3. The Multiple Mini Interview (MMI) is a relatively new interview style which contrasts with traditional panel interviews. Typically, panel interviews were approximately 30 minutes in duration and the interviewee and three to four panel members met in one room. During the traditional panel interview, each panelist took turns asking questions to evaluate different aspects of the candidate's skills and experiences. By contrast, the MMI format consists of multiple stations lasting approximately 8 minutes, each in a different room (physical or virtual), and with a single interviewer for each station.

What do you think are the advantages of the MMI?

What are the disadvantages of the MMI? Why do you think most healthcare professional schools are now using the MMI? Can you prepare for the MMI? Should schools go back to the traditional interview format? What should the qualifications be for MMI interviewers? How many MMI questions should be asked? How are the MMI format and panel format similar?

---

4. A 2016 *JAMA* study examined the prevalence of depression among medical students. The study reported that "the prevalence of depression or depressive symptoms among medical students was 27.2% and that of suicidal ideation was 11.1%." It also concluded that "further research is needed to identify strategies for preventing and treating these disorders in this population."

What further research would you do to explore the issue of depression among medical students?

How do you think these figures compare to the general population? What are some of the risk factors for depression among medical students? Who would you involve in addressing this issue? Should medical schools have screens for depression at the interview stage? Are stress and depression correlated?

Source: https://jamanetwork.com/journals/jama/article-abstract/2589340

5. A 2009 study published in *Pain Research and Management* found that veterinary students received five times more training in pain management than medical students.

Discuss possible reasons for this discrepancy.

What are potential solutions to address this discrepancy? Why do you think the subject of pain could be difficult to incorporate into curriculum in medical schools? Is pain a condition or a symptom? Why is pain an important issue for healthcare providers? Do you think this lack of training among medical students contributed to the opioid crisis?

Source: Watt-Watson J, McGillion M, Hunter J, Choiniere M, Clark AJ, Dewar A, Johnston C, Lynch M, Morley-Forster P, Moulin D, Thie N, von Baeyer CL, Webber K. A survey of prelicensure pain curricula in health science faculties in Canadian universities. *Pain Res Manag*. 2009 Nov-Dec;14(6):439-44. doi: 10.1155/2009/307932. PMID: 20011714; PMCID: PMC2807771.

6. Assessment criteria for entry into most health professional schools include a combination of undergraduate grades, standardized admissions test (e.g., DAT, MCAT, UCAT) results, as well as non-academic criteria (e.g., volunteering).

If you were the Dean of Admissions at your school, would you use the same criteria? Justify your response.

Do you think academic or non-academic criteria are more important? What are some of the drawbacks of standardized tests? How should non-academic criteria be scored? For example, how would you score an Olympic athlete versus someone who raised children during undergraduate school?

---

7. You have been tasked as the social media coordinator for your health professional school of choice.

If you had to choose five photos for the Instagram account for your university's program, which photos would you choose and why?

How did the photos you choose relate to equity, diversity, and inclusion? Which of your photos relates to the role of teamwork in healthcare? Which of your photos relates to the role of resilience in healthcare?

---

8. Before entering into practice, medical students must take the Hippocratic Oath. One of the guiding principles of the oath is "first, do no harm" (or *primum non nocere*, the Latin translation from the original Greek).

If you were to redesign or rephrase the oath, what would it be, and why?

What is the purpose of an oath? How do oaths help to establish ethical standards in various fields? Should all professions be required to take an oath? Is "harm" subjective or objective? How do you see or measure harm as a student or licensed professional? How can professional organizations work to ensure that oaths are updated and revised as needed to reflect evolving ethical standards and best practices? What role do professional oaths play in public trust and confidence in various fields? How do professional oaths intersect with issues of diversity and inclusion, particularly for professionals who may have cultural or religious objections to certain aspects of these oaths?

---

9. Health care can be a challenging profession.

What experiences have you had that lead you to believe you would be well suited to be a healthcare provider?

Why have you chosen this specific health profession versus other healthcare professions? What traits do you think are common to good healthcare providers? What do you think are challenges experienced by healthcare providers?

Chapter 2: Biomedical Topics

10. Compared to traditional medical practices, cosmetic medicine and cosmetic dentistry (e.g., plastic surgery, teeth whitening) present unique ethical issues.

What are some of the ethical issues with cosmetic medicine/dentistry?

With the shortage of general practitioners, should individuals be allowed to practice private cosmetic medicine/dentistry? What are some of the advantages of cosmetic medicine/dentistry? How has social media contributed to the demand for cosmetic medicine and dentistry?

11. Medical tourism is a major global industry whereby individuals travel to other countries with the hopes of accessing medical treatments which may not be available domestically, or which may not be affordable in their home countries. An example of medical tourism is where individuals seek stem cell therapies. For example, one website claims that their stem cell treatment can restore walking function in individuals who are paralyzed from spinal cord injury.

Discuss the ethical issues involved in stem cell tourism.

What is the role of digital literacy in medical tourism? Should countries be required to cover the costs of side effects caused by stem cell tourism? What is the role of hope (or false hope) in these cases? How can healthcare providers and organizations work to educate patients about the risks and potential harms of stem cell tourism or other medical tourism? What are other examples of medical tourism?

---

12. Supervised consumption sites are facilities where individuals can consume drugs under the supervision of trained professionals, in a safe and hygienic environment. These facilities provide a range of services, including harm reduction education, overdose prevention, and referral to health and social services.

What are the potential benefits of supervised consumption sites?

What are potential drawbacks of supervised consumption sites? How can the benefits of supervised consumption sites be measured? Which stakeholders should be involved in the design of supervised consumption sites? Where should supervised consumption sites be located?

---

13. Organ donation is the process of donating organs from a deceased or living person to another person in need of a transplant. Organ donation can save and improve lives, as donated organs can be used to treat a wide range of life-threatening conditions, such as heart and liver failure, kidney disease, and cystic fibrosis. However, there is a shortage of organs available for transplantation, and many people die while waiting for a transplant.

To increase organ donation, should people be allowed to purchase organs for transplant?

What other options are there to increase organ donation rates? How does religious freedom play into organ donation? What screenings should be in place for live organ donation (i.e., when a living person donates organs or part of an organ for transplantation to another person)?

14. In an "opt-out" system for organ donation, the donor's consent is presumed unless there is evidence they did not want to donate. By contrast, "opt-in" systems, as in England and Canada, require donors to give explicit consent while alive.

Discuss the pros and cons of opt-in and opt-out systems for organ donation.

What other options are there to increase organ donation rates? How does religious freedom play into organ donation? How should consent be obtained for organ donation?

15. Labour strikes for healthcare workers can occur when employees in the healthcare industry feel that their wages, benefits, or working conditions are inadequate. Strikes can impact patient care and cause disruptions in hospital and healthcare systems. Labour strikes are on the rise in various healthcare industries.

Should healthcare workers be allowed to strike?

What are the consequences of strikes in health care? Are there alternatives to strikes in health care? Should workers in other sectors such as education or sanitation be allowed to strike?

16. Several roles use the title of "doctor", including medical doctors (MDs), veterinarians, pharmacists, and those with a doctoral degree (PhD). However, some

regulations forbid professionals from using the "doctor" title in a clinical setting, unless they hold a MD degree.

Should nurses with doctoral degrees be referred to as "doctors" in a clinical setting?

How do patients distinguish their healthcare providers? In a healthcare setting, could confusion over who is who cause harm to a patient? What is a "clinical setting"?

---

17. A 2019 Statistics Canada study found that nearly 15 per cent of British Columbians 12 and older have no family doctor (general practitioners, or GPs), a number that is expected to worsen with retiring family doctors. Other regions within and outside of Canada face similar shortages in primary care.

What can governments do to address the shortage of family doctors?

What is the role of foreign medical doctors to address the shortage? Should governments provide higher compensation for family doctors? Should medical schools have specific streams for family doctors? Besides family medicine, what other areas of health care have personnel shortages?

Source: https://www150.statcan.gc.ca/n1/pub/82-625-x/2020001/article/00004-eng.htm

18. The World Health Organization (WHO) defines health as "a state of complete physical, mental, and social well-being, and not merely the absence of disease or infirmity."

Discuss your thoughts on this definition of health.

How can we measure physical well-being? How can we measure mental well-being? How can we measure social well-being? Are physical, mental, and social aspects of health mutually exclusive?

Source: World Health Organization: https://apps.who.int/gb/bd/PDF/bd47/EN/constitution-en.pdf?ua=1

19. Some have defined medicine as "the science and art of healing."

Discuss your thoughts on this definition.

How does science differ from art? How is science similar to art? Should "prevention" be added to this definition? Where does intuition fit into this definition?

20. Evidence-based medicine (EBM) is an approach to clinical practice that uses the best available scientific evidence to guide decision making.

Where do we find "evidence" in various areas of healthcare (e.g., pharmacy, dentistry, medicine, nursing, veterinary medicine, physiotherapy, paramedicine)?

What skills are needed to appraise evidence? What do we do when there is a lack of evidence in health care? What do we do when the evidence is conflicting? How do we know which evidence is the best?

21. More than 10,000 readers of the *BMJ* (formerly *British Medical Journal*) were polled about the most important medical milestone since 1840 (the year the *BMJ* was first published). The majority of readers rated access to clean water and sewage

disposal—"the sanitary revolution"—as the most important medical milestone since 1840. Sanitation was followed by the discovery of antibiotics and the development of anaesthesia for surgical procedures.

In your opinion, what is the greatest medical advance since 1840?

Is sanitation an aspect of medicine? What is the greatest medical advance in just the last 10 years? What advance should be nominated for the next Nobel Prize in medicine?

Source: https://www.ncbi.nlm.nih.gov/pmc/articles/PMC1779856/

---

22. Merriam-Webster explains the historical origins of the term "doctor":

> The English language history of *doctor* starts in the early 14th century, when the word was first applied to a select few who likely knew neither bloodwork nor basketwork. They were equipped for dealing with matters of the soul: they were eminent theologians who had a special seal of approval from the Roman Catholic Church as people able to talk about and explain the doctrines of the Church. They were teachers of a kind, and the word's origin makes this connection. The word *doctor* comes from the Latin word for "teacher," itself from *docēre*, meaning "to teach."

With this definition in mind, do you think the word "doctor" makes sense to apply to medical professionals in the present day?

Are all doctors involved in teaching? What label could replace "doctor"?

Source: https://www.merriam-webster.com/words-at-play/the-history-of-doctor

23. Several questionable internet websites have claimed that artificial sweeteners, particularly aspartame, can cause neurological diseases such as multiple sclerosis. Their explanation is that when ingested, aspartame is broken down into its components: aspartic acid, phenylalanine, and methanol. Methanol can further metabolize into formic acid, which, in large quantities, may be harmful. Formic acid is in the same class of drugs including cyanide and arsenic. Formic acid is also the key component responsible for the painful sting of fire ants. When fire ants bite, they inject venom containing formic acid into their victim's skin, causing an intense burning sensation and skin irritation.

What are your thoughts on this claim?

Discuss another example of a faulty claim you have seen on the internet. What role does social media play in propagating false claims? What skills do the general public need to analyze such claims?

24. The Canadian Blood Services recently partnered with private companies that pay people for donating their blood plasma for use in life-saving transfusions after blood loss. Other countries have adopted similar partnerships.

Discuss the ethical issues involved in paying people for blood donations.

What are other options to increase blood donation rates? Should eligibility criteria for blood donations be relaxed?

---

25. In the late 1990s, a Texas jury attributed the death of a 42-year-old patient to an illegible prescription. The physician wrote a prescription for 20 mg isordil, an antianginal drug, which was misread by the pharmacist as 20 mg Plendil, an antihypertensive drug. After taking the incorrect medication, the patient had a heart attack and died days later.

How do you think medical errors like this could be avoided?

What are other examples of medical errors? To whom should medical errors be disclosed? When should medical errors be disclosed? Does admitting fault increase the probability of a lawsuit?

Source: Charatan F. Compensation awarded for death after illegible prescription. *West J Med.* 2000 Feb;172(2):80. doi: 10.1136/ewjm.172.2.80. PMID: 10693362; PMCID: PMC1070756.

---

26. The *Canada Health Act* is Canada's federal legislation for publicly funded healthcare insurance. It includes five major principles: public administration, accessibility, comprehensiveness, universality, and portability. Other countries with publicly funded health care adopted similar principles.

Discuss what each of these terms means to you.

Are there any terms you would add to the list of principles, as a policy maker? Which of these terms is the most important? Which term is the least important? Which of these terms most relates to equity and diversity? Do all these terms apply to healthcare systems in other countries?

27. You are the acting President/Prime Minister for your country.

What would you advocate for as the top priority for healthcare in your nation?

How did you decide on the top priority within the health sector? How would you decide on priorities across sectors (e.g., education, environment, health)? How did the COVID-19 pandemic change priorities for health care?

28. The news is full of stories about overcrowded emergency rooms and long wait times for procedures. To address this issue, some experts have suggested the use of "deterrent fees", a small charge anyone would have to pay when they initiate contact with a primary care provider.

What are the benefits of deterrent fees?

What are the downsides of deterrent fees? Would a deterrent fee deter you from accessing a healthcare provider? What groups could be further disadvantaged from deterrent fees?

29. In many areas, if you call 911 for an ambulance, you are responsible for covering a portion of the cost. If you do not have supplemental insurance, you are responsible for paying the full cost of the ambulance.

Discuss the implications of this ambulance user fee.

What are the downsides of ambulance user fees? Would this fee deter you from calling 911? What groups could be further disadvantaged from ambulance user fees? How do ambulance user fees intersect with other issues of healthcare financing and access, and how can they be effectively integrated into broader discussions around healthcare reform?

---

30. Some studies have shown that physicians die by suicide more frequently than non-physicians.

Why do you think this is the case?

Who do you think should be involved in helping to solve this problem? Do you think the COVID-19 pandemic compounded this problem? Could medical schools and their admission criteria help prevent this problem? Do you think other healthcare professions show similar trends?

---

31. Read the following excerpt from a commentary written by a healthcare provider, entitled "Shades of grey: patient versus client":

> "I came here for the cut on the head and see no reason to do what you say. You are here to serve me. My taxes pay your salary." This angry outburst came from a parent when one of our emergency nurses asked her to remove her baby's clothes so she could check the child's heart rate. As I listened to the client demanding the service she wanted, I wondered if we had pushed this client concept a little too far. The baby

is our patient-client, and the mother is the decision-maker client who pays our salaries. Where do we draw the line?

Discuss your thoughts on this scenario.

Who is the client in this case? Should patients be referred to as "clients"? Why did the healthcare provider title this commentary "shades of grey"?

Source: *Canadian Medical Association Journal*: https://www.cmaj.ca/content/180/4/472

32. Read the following case from the College of Family Physicians of Canada (CFPC):

> Mr. B is a 37-year-old male patient of yours with a long history of schizophrenia and, more recently, end-stage idiopathic cardiomyopathy who has been refused consideration for the cardiac transplantation waiting list. Mr. B has been unemployed for a number of years due to his illnesses. He is on maximal medications for his heart disease and continues to decline. Mr. B's psychiatric condition is currently under control with an expensive new oral neuroleptic. The Government Assisted Drug Plan has recently been overhauled and may no longer cover this drug.

What do you think would be the responsibilities of the healthcare provider in this case?

Should healthcare providers be required to take on the role of advocates? Should the government fund this new drug? How could social stigma around mental illness play into priority setting for drug funding?

Source: https://www.cfpc.ca/CFPC/media/Resources/Ethics/bioethics_en.pdf

---

33. You have been tasked with teaching an ethics seminar for a kindergarten classroom.

How would you explain the four main principles of biomedical ethics (beneficence, non maleficence, autonomy, and justice) to a kindergarten (age 5 to 6) audience?

How would you prepare for the seminar? What language would you be careful to avoid in your seminar? How would you maintain interest in children with short attention spans? How would you relate these principles to everyday life for a kindergartener?

---

34. According to the *British Medical Bulletin*: "Triage refers to situations of emergency where different patient priority groups are established in order for scarce vital health resources to be distributed. Today, triage is mainly used in healthcare contexts during disasters, mass casualty incidents, in emergency departments and ICU."

What criteria should be used to establish priority in these situations?

What are some of the ethical and legal considerations around triage, particularly in cases where resources are limited and difficult decisions must be made about allocating care? How does triage differ in different healthcare settings, such as in the emergency department versus in disaster-response scenarios? How can healthcare providers work to establish clear and transparent protocols for triage in order to ensure fair and consistent decision-making?

Source: Vinay R, Baumann H, Biller-Andorno N. Ethics of ICU triage during COVID-19. *Br Med Bull.* 2021 Jun 10;138(1):5-15. doi: 10.1093/bmb/ldab009. PMID: 34057458; PMCID: PMC8195142.

---

35. A parent brings their infant to a community health clinic for a six-month routine check-up. The parent is unsure about having the infant vaccinated, particularly over concerns about the additives in the vaccines.

How would you handle this case?

How can healthcare providers approach conversations with families who are hesitant or resistant to receiving vaccines, and what are some effective strategies for addressing their concerns? How can healthcare providers counteract misinformation about vaccines, particularly in online and social media spaces? How can healthcare providers address vaccine hesitancy or refusal in a culturally sensitive and respectful way, particularly in communities where there may be distrust or historical trauma around medical interventions?

---

36. A *JAMA* study found that "women physicians were significantly more likely to report not working full-time than men physicians, and differences were even greater among women with children compared with men with children."

Discuss your thoughts on this finding.

What are some of the factors contributing to gender disparities in medical professions, and how do these disparities impact the overall quality of care and patient outcomes? How can healthcare organizations and professional associations work to address gender disparities in medical education and training, including through mentorship? What are some of the specific challenges faced by women in medical professions, such as bias, discrimination, and work-life balance, and how can these challenges be effectively addressed? How do gender disparities in medical professions intersect with other forms of marginalization and inequity, such as those based on race, ethnicity, or socioeconomic status?

Source: Frank E, Zhao Z, Sen S, Guille C. Gender Disparities in Work and Parental Status Among Early Career Physicians. *JAMA Netw Open.* 2019;2(8):e198340. doi:10.1001/jamanetworkopen.2019.8340

---

37. It is important for patients to feel comfortable and respected during their healthcare experience, and the option to select a healthcare provider based on their gender preference could contribute to that.

Should patients be able to select the gender of their healthcare providers?

Under what circumstances would this be possible? Under what circumstances would this not be possible? Does this apply to other factors such as race or age?

---

38. It is important for patients to feel comfortable and respected during their healthcare experience, and the option to select a healthcare provider based on racial background could contribute to that.

Should patients be able to select the race of their healthcare providers?

Under what circumstances would this be possible? Under what circumstances would this not be possible? Does this apply to other factors such as sex or age?

39. Read the case below from the Canadian Medical Protective Association (CMPA):

> You are the physician for two children, ages four and six, who live with their mother and visit their father on weekends and some evenings. The parents communicate poorly, and the father feels his ex-wife has not kept him informed about the children's health and medications. He has requested a copy of their records. The mother would prefer that her ex-husband not be given any information because she believes he will misuse the information in an attempt to obtain custody. The children have had ear infections. Both are allergic to cats. There is a cat in the mother's home.

What ethical issues does this case raise?

Is the father entitled to medical information about his children? How would you handle the conflict between the parents? How would you make sure the children are receiving optimal care?

Source: https://www.cmpa-acpm.ca/en/advice-publications/browse-articles/2005/responding-to-requests-for-children-s-medical-records

40. Read the case below from the CFPC Ethics Committee:

> You have finally convinced your skeptical patient to take a low dose medication for her poorly controlled hypertension. At the pharmacy the patient receives, along with her pill, a printed list of all the potential side effects of hydrochlorthiazide. At the next visit your patient informs you that she has not taken the medication, nor will she because

of "potential dangerous side effects," and shows you the long list provided by the pharmacist.

Is the pharmacist wrong in providing this information to the patient?

How can the physician and pharmacist work together in this case to provide optimal care? How could the healthcare providers gauge if the side effects were the only reason the patient wasn't taking their medication? How can healthcare providers educate patients to better understand health information? Do all patients want the same amount of information?

Source: https://www.cfpc.ca/CFPC/media/Resources/Ethics/bioethics_en.pdf

---

41. Read the case below from the Canadian Medical Protective Association (CMPA):

> A 16-year-old saw a family physician for symptoms of severe depression. Speaking with the patient, the physician determined that the patient was mature enough to understand the seriousness of his symptoms and the need to address them. The patient also requested that the physician not consult with his parents. The physician referred the patient to an adolescent day treatment program where he was followed by a psychiatrist. He was diagnosed with major depression and agoraphobia. After learning her child was undergoing treatment, the patient's mother filed a College complaint, alleging that the family physician and psychiatrist did not obtain proper consent for her child to attend the treatment program. The College stated that, by complying with the patient's request not to consult his parents, both

physicians acted in the best interests of the patient and in accordance to College practice.

Discuss your thoughts on this case.

Did the governing body make the right decision? Should age be the major factor in deciding competency? Going forward, how can the physician help repair their relationship with the parents?

Source: https://www.cmpa-acpm.ca/en/advice-publications/browse-articles/2014/can-a-child-provide-consent

---

42. Read the case below from the Royal College of Physicians and Surgeons of Canada (written by Philip Hebert, MD):

> A hospital-based obstetrician–gynecologist does a Pap smear on a 48-year-old patient ... Shortly after doing the Pap smear and before the report on it came back, the gynecologist leaves the country for an extended period to work overseas. When he returns 12 months later, he sees the patient again for her annual visit in his private office and is surprised to discover that she has an advanced form of cervical cancer. He realizes that last year's Pap report is not in her file. Calling the lab, he is told, to his great dismay, that the smear showed evidence of pre-cancerous cells. Having no replacement for him when he had left a year ago, the hospital had closed the clinic in which he had worked. Unfortunately, no arrangements were made to handle his reports. As a consequence, the Pap smear report for the

> patient was unseen by anyone, and she was not told about her medical condition nor did she receive treatment for it.

What ought the clinician to do or say, if anything, to the patient?

What are the ethical and legal implications of disclosing medical errors to patients, and what are some potential benefits and drawbacks of doing so? How can healthcare providers and organizations establish effective protocols for disclosing medical errors, and what are some best practices for communicating with patients and families in these situations? How can patients be empowered to play a more active role in their own healthcare, including by encouraging and facilitating transparency and disclosure of medical errors? How can healthcare organizations and providers work to create a culture of transparency and accountability around medical errors in order to improve patient safety and overall quality of care?

Source: www.royalcollege.ca: 2.1.4 "The Missed Result"

---

43. Read the following case below (written by Bryan S. Vartabedian, MD):

> After a long day at work, Dr. Baker sits down to check her email and finds a forward from an old medical school friend. "I thought you'd enjoy this," her friend has written. The link takes her to a blog called "theGrouchyMD: musings of an overworked Texas OB/GYN resident." The first posts tells the story of "Jane," a 53- year-old woman trying to get pregnant. The blogger expresses the opinion that "Jane may be well intentioned, but I can't help thinking that what she's doing is selfish and irresponsible" and posts some links to news articles on uses and misuses of reproductive technology. A lively debate follows among blog readers, who identify

themselves as members of the public, medical students, and other physicians. Earlier posts are on topics ranging from health care reform to "war stories" ("Caught a twin vaginal delivery today," reads one post, "textbook! Now that's real obstetrics!"). Some posts are centered on readers' questions, like one explaining the differences between some types of oral contraceptive pills.

As she scrolls down, Dr. Baker becomes increasingly concerned; the patients begin to sound familiar, as do "Dr. B" and "Dr. H," theGrouchyMD's colleagues. The blogger apparently not only practices in the same hospital as Dr. Baker does, but appears to be in the same program. The next day, Dr. Baker confronts Dr. O'Connell, her fellow third-year resident. "You're theGrouchyMD, right?" she starts. "I'm concerned about what you're doing. I know you changed the names, but what if someone recognizes herself?"

Do you share Dr. Baker's concerns?

How do the issues of consent and privacy relate to this case? What are the potential harms to the patient? Are there any benefits of blogs written by healthcare professionals? What safeguards can be used to ensure privacy in these cases?

Source: https://journalofethics.ama-assn.org/sites/journalofethics.ama-assn.org/files/2018-06/vm-1107.pdf

44. According to the *BMJ Opinion*:

> The COVID-19 pandemic has had a dramatic global health impact particularly in countries which are demographically older and with populations carrying significant co-morbidities. While the focus has been on direct mortality caused by COVID-19, there has been little attention paid to the indirect impact on other significant health conditions, such as cancer, caused by national measures to contain the SARS-Cov-2 virus spread, so-called nonpharmaceutical interventions (NPI). NPI's encompass a wide range of measures including "hard lockdowns" and there is a growing realization that particularly national or regional lockdowns, come at a cost to wider health and welfare of societies.

What are some of the short- and long-term health consequences of the COVID-19 pandemic, both for individuals who have contracted the virus and for the general population?

How has the pandemic impacted mental health and well-being, and what are some effective strategies for addressing these challenges? What are some of the ways in which the pandemic has exacerbated existing health disparities and inequities, particularly among marginalized and underserved populations? How has the pandemic affected access to health care and health services, and what are some potential long-term consequences of these disruptions? How can policy makers and healthcare professionals work to mitigate the health consequences of the pandemic, both through immediate response efforts and through longer-term public health interventions and policy changes?

Source: https://blogs.bmj.com/bmj/2020/11/05/counting-the-invisible-costs-o
f-covid-19-the-cancer-pandemic/

45. Social determinants of health refer to the social, economic, and environmental conditions that can significantly impact an individual's overall health and well-being.

What are specific examples of social determinants of health, and how do they impact an individual's overall health and well-being?

How do factors such as income, education, and employment status influence an individual's access to health care and ability to lead a healthy lifestyle? How do race, ethnicity, and cultural factors impact an individual's experience of health care and health outcomes? What are some examples of policies and interventions aimed at addressing social determinants of health, and how effective have these programs been? How are the social determinants of health linked to the social determinants of criminal behaviour?

46. Rural areas of Canada, and other countries, face a shortage of healthcare professionals. The Government of Canada has invested in programs to encourage healthcare professional students to train in rural areas.

What are some of the underlying causes of shortages of healthcare professionals in rural locations?

How can the shortage of doctors in rural areas impact the health and well-being of local communities? What are some current efforts to address rural doctor shortages and how effective have these initiatives been? How can rural doctor shortages be addressed in a way that takes into account the unique challenges and opportunities of rural communities, such as geographic isolation or cultural factors? What can be done to ensure that all individuals have access to high-quality medical care, regardless of where they live?

---

47. Read the following case from the CFPC Ethics Committee:

> A close friend wants to become your patient. You've been told this is not a good idea and initially refused him. He says he can talk to you about issues he would find difficult talking to others about. You also "know his background far better than some stranger". He urges you to reconsider.

What are the ethical issues in this case?

What action would you take if you were a health care provider in this case? How would your decision be affected if you lived in a rural area?

Source: https://www.cfpc.ca/CFPC/media/Resources/Ethics/bioethics_en.pdf

---

48. Read the following case from the CFPC Ethics Committee:

> You are examining Jillian M, a 10-month-old baby, in your office while her parents are briefly out of the office. Momentarily distracted, you allow the baby to fall off the examining table. Although crying, the child seems unharmed.

What ought the healthcare provider say to the parents who were not in the room at the time?

To whom should an error such as this be disclosed, in addition to the parents? Should all errors be disclosed? What is the potential harm of disclosing errors? What are the benefits of disclosing errors? What safeguards could be put in place to avoid this type of error?

Source: https://www.cfpc.ca/CFPC/media/Resources/Ethics/bioethics_en.pdf

49. Imagine that you are a healthcare provider, working in a hospital where a patient with an extremely rare disease is being treated.

How would you go about researching the rare disease to provide the best possible care for the patient?

What would you do if you were not able to find any research on this case? How would you handle researching this case with your heavy workload for your other patients? How could you advocate for more research on this rare disease?

50. Read the following passage from the Centers for Disease Control and Prevention (CDC):

> During the early years of the HIV epidemic, many states implemented HIV-specific criminal exposure laws to discourage actions that might lead to transmission, promote safer sex practices, and, in some cases, receive funds to support HIV prevention activities. These laws were passed

at a time when little was known about HIV including how HIV was transmitted and how best to treat the virus. Many of these state laws, then and now, criminalize actions that cannot transmit HIV—such as biting or spitting—and apply regardless of actual transmission, or intent. After more than 40 years of HIV research and significant biomedical advancements to treat and prevent HIV transmission, many state laws are now outdated and do not reflect our current understanding of HIV. In many cases, this same standard is not applied to other infectious, treatable diseases. Further, these laws have been shown to increase stigma, exacerbate disparities, and may discourage HIV testing.

Discuss your thoughts on this passage.

Why would some of these laws discourage HIV testing? Are there other cases where laws lag behind the most current evidence?

Source: https://www.cdc.gov/hiv/policies/law/states/exposure.html

---

51. International surrogacy is a process in which intended parents from one country travel to another country to hire a surrogate mother to carry their child. This is often done when surrogacy is either illegal or too expensive in their home country. International surrogacy raises several ethical and legal issues. Recently, governments in India and other countries have banned surrogate services for foreigners.

Discuss the ethical implications of international surrogacy.

What are potential benefits of international surrogacy? What are potential drawbacks of international surrogacy? Who should be involved in deciding regulations or laws for international surrogacy?

52. Read the following case from the CFPC Ethics Committee:

> The provincial College of Physicians and Surgeons has just sent all doctors a newsletter on the subject of receiving gifts from patients. In it, the registrar states, "patients like to show their appreciation with gifts. However, if the gift is more substantial than a hand tatted doily, the physician will have an ethical problem." You have just assisted at another successful delivery of a healthy baby. The delighted and grateful parents give you a gift of: a) a bottle of single malt whiskey worth $60-; OR b) a $100- bill; OR c) use of their resort condo at Whistler for a weekend.

Is there an ethical difference in the different gifts?

Would it be different if the patients were rich or poor? What is too big a gift? Should the governing bodies of health care professions set official policies on gifts? What should be the consequences of healthcare providers who violate this policy? Do the practice and norms of gift giving vary in different cultures?

Source: https://www.cfpc.ca/CFPC/media/Resources/Ethics/bioethics_en.pdf

53. Major changes have recently come to several healthcare systems to alleviate the strain on family doctors. For example, pharmacists will be able to administer

more vaccines and renew prescriptions, and they will be able to prescribe drugs for conditions such as urinary tract infections, allergies, acne, and indigestion, as well as contraception.

What considerations should be taken into account when expanding the scope of pharmacy practice?

What is the role of pharmacists in your region? How should the scope of prescribing be determined? What benefits or drawbacks could arise with respect to patient safety and access to care? Could any conflicts of interest arise from the increased involvement of pharmacists in prescribing?

---

54. A quick internet search of celebrities with cancer brings up articles with headlines such as "Celebs Who Battled Cancer and Won!" Similarly, articles about dying from cancer often use the phrase: "they lost their battle" to cancer.

How could the battle metaphor be problematic for patients with cancer?

How could the battle metaphor be problematic for healthcare providers? Could the battle metaphor be empowering in any ways? What other metaphor could be used to portray a difficult health journey?

Chapter 3: Pharmaceutical Topics

55. Aducanumab is a monoclonal antibody designed to target beta-amyloid, a protein that accumulates in the brains of Alzheimer's patients and is believed to contribute to the disease's progression. While initial clinical trials showed promise, subsequent studies raised concerns about the drug's efficacy, and the FDA granted conditional approval in 2021 based on uncertain evidence.

Should the FDA have granted conditional approval of this drug?

What information should factor into whether or not a drug is approved? Often, regulatory agencies take years to approve a drug; should this process be expedited, and under what conditions? Should individuals with Alzheimer's be allowed to consent to the use of this drug?

56. Direct-to-consumer (DTC) advertising for medications is a type of marketing strategy by which pharmaceutical companies promote their prescription drugs directly to consumers through various channels, such as television commercials, print ads, social media, and websites. This type of advertising is typically targeted at patients who may have a medical condition that can be treated with the advertised drug.

Why is DTC controversial?

What are the ethical concerns surrounding DTC advertising from pharmaceutical companies, particularly in terms of the potential for these ads to overemphasize the benefits of medications while downplaying or ignoring potential side effects? What information should be provided in television commercials to patients? What percentage of a pharmaceutical company's budget should be spent on DTC? What role do healthcare providers play in responding to DTC advertising, and how can they work to ensure that patients are making informed decisions about their treatment options?

---

57. You are a healthcare provider who is involved in a clinical trial of a new drug. Of your 16 patients enrolled in the trial, 13 developed moderately severe rashes. When you report this finding to the pharmaceutical company, they don't seem interested in your concerns.

How would you deal with this problem?

What would you do if you felt your concerns weren't being listened to? What would you do if the "moderately severe" rashes developed into "very severe" rashes? How would your response differ if you worked for the company versus a government institution such as a university?

Chapter 4: Academic/Research Topics

58. Scientific conferences play a crucial role in the lives of researchers. They provide an opportunity for scientists to exchange ideas, forge research collaborations, establish connections with funding agencies, and attract new members to their research programs. Unfortunately, early-career researchers often encounter barriers to attending conferences, particularly when it comes to childcare. This presents a major challenge for parent-researchers who must balance attending key conference events with finding suitable care for their children.

Should academic conferences be obligated to provide childcare for parents attending their events?

If a conference does not provide childcare, how will this affect the diversity of conference attendees? Would you be willing to cover the costs of childcare for other conference attendees? If you were a conference organizer, what concerns would you have about providing childcare at the conference?

59. Scientific conferences play a crucial role in advancing knowledge across all fields of study. Traditionally, these conferences have been conducted in person, requiring attendees to travel and contribute to carbon emissions through air travel. However, with the recognition of the negative impact on the environment, the value

of digitizing conferences through synchronous and/or asynchronous internet-based techniques has been emphasized.

If you were planning a scientific conference, would you choose an in-person or digital format? Justify your choice.

What are some of the ethical considerations around the use of university funds to support conference travel, particularly considering competing budget priorities and concerns around financial transparency and accountability? Who should provide funding for scientists to attend scientific conferences? In 20 years, do you think in-person scientific conferences will still exist?

---

60. The American Association for the Advancement of Science (AAAS) recently released a policy regarding artificial intelligence (AI) and authorship of scientific studies. The policy states that

> an AI program cannot be an author of a *Science* journal paper. Text generated from AI, machine learning, or similar algorithmic tools cannot be used in papers published in *Science* journals, nor can the accompanying figures, images, or graphics be the products of such tools, without explicit permission from the editors.

Do you think that a violation of this policy constitutes scientific misconduct?

Should you be allowed to use AI if you cite AI as an author? Do you think AI will result in a general loss of writing skills? Should students be able to use AI in classes? How could instructors incorporate AI into the university classroom?

Source: https://www.science.org/content/page/science-journals-editorial-policies

---

61. The Nobel Prize is one of the most prestigious and recognized awards in the world, awarded annually for outstanding contributions in the fields of physics, chemistry, medicine, literature, peace, and economics. Established by the Swedish inventor and industrialist Alfred Nobel in 1895, the prize is awarded based on the recommendations of committees appointed by various institutions, including the Royal Swedish Academy of Sciences and the Norwegian Nobel Committee. The winners are chosen based on their significant achievements and discoveries that have had a profound impact on human progress, and their work is recognized as having made a major contribution to the advancement of knowledge and the betterment of society. The Nobel Prize is not only a celebration of scientific and intellectual achievement, but also serves as a reminder of the importance of innovation, dedication, and perseverance in shaping the world we live in today.

The Nobel Prize committee has appointed you to pick a new category for the Nobel Prize. What category would you choose? Justify your response.

Are there any Nobel Prize categories you would remove? How do you think "profound impact on human progress" can be judged? Who should be on the selection committee for the Nobel Prize? Should the Nobel Prize be awarded to just one person, or should more than one person be eligible to win a given prize?

62. In health research, there is a growing recognition of the importance of involving patients or the general public as partners in various stages of the research process.

What are the potential benefits of involving patients in research?

What are the potential drawbacks of involving patients in research? In addition to scientists and patients, who else should be involved in conducting research and determining research priorities? For rare diseases, do you think, in some cases, patients know more about a health condition than their trained healthcare providers?

---

63. Healthcare professionals and researchers have a responsibility to inform patients before carrying out experimental treatment interventions as part of clinical research. This can be particularly difficult in emergency situations such as traumatic brain injury and stroke. These individuals, often presenting with acute cognitive impairment, may be unable to provide informed consent in a timely fashion.

Should investigators be allowed to enrol subjects into research studies in emergency situations?

What should patients be informed about? How should patients be informed? Who should inform patients?

---

64. The "democratization of science" refers to the idea that scientific knowledge and research should be accessible, transparent, and inclusive to all members of society, regardless of their background, education level, or socioeconomic status. This includes promoting open access to scientific publications, encouraging citizen science participation, increasing diversity and inclusivity in scientific research, and promoting public engagement with science. By democratizing science, the hope

is to improve the quality and relevance of scientific research, foster innovation, and promote social justice by ensuring that scientific knowledge is used for the betterment of all.

What role do scientists have in the democratization of science?

How does the democratization of science challenge traditional models of scientific research and knowledge dissemination? What role do healthcare providers have in the democratization of science? What role does the general public have in the democratization of science? Do you think the democratization of science will improve the quality of scientific research?

---

65. According to the Government of Canada,

> an "incidental finding" is a discovery about research participants or prospective participants that is made in the course of research, <u>but is outside the objectives of the research study</u>.... Rapid technological advances, the evolution of research capabilities, large volumes of data, and the push for innovation contribute to increasing the probability of incidental findings in research involving humans. Examples of incidental findings that may be considered material include: an unexpected mass or vascular abnormality on a computed tomography (CT) scan or a magnetic resonance imaging (MRI); a genome sequence that reveals additional genetic variation for a participant, such as high risk for cancer; and a discovery of physical abuse or suicidality in studies unrelated to those phenomena.

Should researchers be required to disclose incidental findings?

Who should be responsible for disclosing the incidental finding to the research participants? If you were a participant in a research study, would you want to know about incidental findings? Could disclosing of incidental findings impact participation in future research? How does technological innovation contribute to the likelihood of incidental findings? Should incidental findings be included in consent forms prior to participation in research?

Source: https://ethics.gc.ca/eng/incidental_findings.html

———•———

66. High school science fairs provide a platform for students to showcase their creativity and scientific skills. These fairs encourage students to think outside the box and come up with innovative ideas to solve real-world problems. When the parents of children who participate in science fairs are scientists, they may have an unfair advantage in helping their children with their projects.

How can science fair organizations combat this issue?

Why are science fairs an important part of high school education? Do you think science fairs are representative of the general population? Which stakeholders can ensure more fair representation in science fairs? In addition to scientific knowledge, what other benefits do science fairs have for youth?

Chapter 5: Equity, Diversity, and Inclusion (EDI) and Cultural Sensitivity Topics

67. Equity, diversity, and inclusion (EDI) are critical concepts for promoting fairness and social justice in society.

Please share what the terms diversity, equity, and inclusion (EDI) mean to you and why they are important.

How does inclusion differ from equity? Is equity the same as equality? How can health professional schools ensure EDI principles are included in training? Which other sectors of society could benefit from EDI training?

68. Land acknowledgements are statements that recognize and honour the Indigenous Peoples who inhabited a particular land before colonization. They are typically read aloud at the beginning of events, meetings, or ceremonies, and serve

to acknowledge the ongoing presence and contributions of Indigenous communities. According to CBC News regarding land acknowledgements:

> They've become so commonplace that you'll hear land acknowledgments at the start of hockey games, during academic conferences and even written at the bottom of corporate email signatures. In an era of reconciliation, they're political statements meant to recognize First Nations, Inuit, and Metis territory, however many Indigenous people argue they've grown to become superficial, performative—and problematic.

What are some of the potential benefits of land acknowledgements?

How do land acknowledgements relate to "truth and reconciliation"? Should universities be required to include land acknowledgements in the classroom setting? What are some of the potential drawbacks of land acknowledgements?

Source: https://www.cbc.ca/news/indigenous/land-acknowledgments-what-s-wrong-with-them-1.6217931

---

69. The term "decolonize" is defined as "to free from the dominating influence of a colonizing power." Decolonization is the process of undoing the effects of colonization. It involves challenging the structures of power, knowledge, and representation that were imposed by colonialism, and empowering Indigenous Peoples to reclaim their cultural, political, and economic sovereignty. Decolonization is an ongoing and complex process that involves confronting the ongoing impacts of colonialism, such as land dispossession, cultural erasure, and systemic discrimination.

Discuss how decolonization is relevant to the past, present, and future where you live.

What are some of the harmful influences of colonizing powers? Why is decolonization ongoing? Why is decolonization complex? How long do you think the process of decolonization should take?

Source: Merriam-Webster.com

---

70. Indigenization is a process that focuses on incorporating Indigenous knowledges, perspectives, and practices into the structures of institutions.

Why is indigenization an important focus for universities?

How does indigenization relate to truth and reconciliation? How does indigenization relate to equity, diversity, and inclusion (EDI)? How can universities indigenize courses? Who should be involved in indigenization at universities?

---

71. The National Day of Truth and Reconciliation is a day of remembrance and reflection in Canada that honours the Indigenous children who were taken from their families and forced to attend residential schools. The day is held on September 30th and encourages Canadians to reflect on the ongoing legacy of residential schools and the impact they have had on Indigenous Peoples. September 30 is also Orange Shirt Day, which honours the story of Phyllis Webstad, a former residential school student who had her orange shirt taken away on her first day at a residential school. The day was made a federal statutory holiday last year following a recommendation made by the Truth and Reconciliation Commission.

What do the terms "truth" and "reconciliation" mean to you in this context?

Why is it important to recognize residential schools with a national day of recognition? Should this day replace other days of recognition like Remembrance Day? In which grade should youth be first taught about residential schools?

―――●―――

72. The legacy of slavery in the United States is a complex and ongoing issue that has shaped the nation's history and continues to impact society today. Slavery was a brutal institution that involved the forced labour and exploitation of millions of Black people over hundreds of years. As the United States confronts its legacy of slavery and systemic racism, many have demanded the removal of monuments (e.g., statues) honouring the Confederacy, a political group of eleven states whose agricultural economy was dependent on Black slave labour.

What should we do with these types of monuments in the present day?

Should the statues be removed? What alternatives to removal are there? Why is the legacy of slavery described as "complex and ongoing"? Should schools or street names, which also honour the Confederacy, be changed?

―――●―――

73. "Cultural appropriation" refers to the use of objects or elements of a non-dominant culture in a way that reinforces stereotypes and doesn't respect their original meaning. It also includes the unauthorized use of parts of a culture (dress, dance, etc.).

How does cultural appropriation differ from cultural appreciation?

Does intent matter in the context of cultural appropriation? Who can give permission to authorize the use of parts of a culture?

Chapter 6: General University Topics

74. Pass/fail university classes operate on a binary grading system, meaning that no letter grade is recorded on the university transcript. Universities typically award passes for students earning any grade higher than a D.

Discuss the pros and cons of a pass/fail grading system.

Do you think a pass/fail grading system discourages competition? Do you think a pass/fail grading system just redistributes pressure to other areas such as extra-curricular activities? How does a pass/fail grading system encourage learning? What percentage should be a cut-off for a pass/fail grading system?

75. In an opinion piece, Alex Usher comments on the topic of Canada's growing reliance on international students at universities. He notes:

> Fees for international students, which average about four times what domestic students pay, now equal 12 percent of operating revenue and 35 percent of all fees collected by institutions, and these proportions continue to climb. Already some major institutions, including the University of Toronto, are receiving more money from international

students than they get in operating grants from their provincial governments.

Discuss the ethical implications of this reliance on international student fees.

Is this reliance on international student fees sustainable? What other solutions could universities use to increase funding? Do you think universities should spend resources on advertising for international students?

Source: https://policyoptions.irpp.org/magazines/august-2018/canadas-growing-reliance-on-international-students/

---

76. University students are faced with high textbook costs and with the pressure to purchase them to avoid poor grades. In some university courses, students are required to purchase textbooks written by their professors.

Discuss how this could present a conflict of interest.

Did you ever have to purchase a textbook that was authored by one of your professors? If so, how did you feel? If not, discuss how you would feel if you were in this situation. Should all university textbooks be free and open access?

---

77. You are a first-year student preparing for an anatomy exam. One of your classmates just emailed you a file containing exams from past years.

Discuss your actions in this scenario.

Would you contact your professor? Would you share the file? Should a student who shares the exam file fail the course?

78. According to CBC News, an "investigation has revealed how fast and convenient it is to purchase a custom academic essay that can go undetected by university professors and plagiarism software. Academic integrity experts say businesses that sell custom-written papers are proliferating across the country." You are invited by the university to be part of a focus group to combat this issue.

What solutions would you offer?

Who should be included in the focus group? What are some of the consequences of plagiarism, both for students and for academic institutions? What should be the consequences of breaches of academic integrity? What are some of the challenges associated with detecting and addressing plagiarism, particularly in an age of digital content and online resources?

Source: https: https://www.cbc.ca/news/canada/toronto/contract-cheating-nursing-investigation-1.5109322

79. A university Dean is a high-ranking official responsible for overseeing an academic unit. The Dean's role involves managing faculty, budgeting, curriculum development, fundraising, program evaluation, and student affairs. Deans are also responsible for setting a vision and mission statement for their academic unit. You are the student representative on the hiring committee for selecting a new Dean in the Faculty of Health Sciences at a major university.

What selection criteria would you use for hiring a new Dean?

In addition to you, who else should serve on the selection committee? How would you incorporate principles of equity, diversity, and inclusion (EDI) into the hiring process? How do the criteria for hiring a Dean differ from the criteria for hiring a new professor?

---

80. You are a first-year university student working on a group project with three other students. One of your group members, Larry, has been late for every group meeting, and when he does show up, he doesn't contribute to any of the discussion or written work.

How would you handle this situation?

Would you approach the instructor about this problem? Should Larry receive the same mark as all others in the group? Should university courses include group work? What skills can be learned from group work?

Chapter 7: Media & Technology Topics

81. Fact checking is an essential practice in today's information age. A fact checker is a professional who makes sure the information their organization provides is accurate and true. They can work both in print and broadcast media formats.

Why is the role of fact checkers important?

What skills do you think a fact checker requires? What sources do fact checkers use to confirm the veracity of claims? Do you think fact checkers can be biased? Can artificial intelligence (AI) fact check? In what area could you serve as a fact checker? What role do fact checkers play in the spread of misinformation? Are all fact checkers credible?

82. The Stanford History Education Group (SHEG) is an educational research group that focuses on improving students' ability to evaluate digital information critically. The group has developed several resources and tools that help students evaluate the credibility of online sources, including the "Civic Online Reasoning" curriculum, which teaches students how to identify misleading information and evaluate the credibility of sources. The SHEG also created the "Sourcing the News" website, which provides tools for evaluating news articles and helps users identify bias and misinformation. In November 2016, SHEG released the results of a study showing that youth lacked basic skills of digital evaluation. For example, they found that "Fifty-two percent of students

believed a grainy video claiming to show ballot stuffing in the 2016 Democratic primaries (the video was actually shot in Russia) constituted 'strong evidence' of voter fraud in the US" and "[a]mong more than 3,000 responses, only three students tracked down the source of the video, even though a quick search turns up a variety of articles exposing the ruse."

What are your concerns about this finding from SHEG?

What is digital information? What strategies do you use to evaluate digital information? What are the consequences of having low digital information literacy? What is the role of digital information in a healthcare field?

Source: https://sheg.stanford.edu/students-civic-online-reasoning

---

83. Celebrities have long been known to leverage their fame for various purposes. Some celebrities use this fame to launch their own businesses, products, or services, capitalizing on their popularity and appeal to their fan base. This "side hustle" may involve promoting or selling products such as cosmetics and alcohol.

What ethical obligations do celebrities have when leveraging their extreme fame and influence for personal gain?

How do these obligations differ from those of individuals without such a high level of notoriety? Has social media contributed to this issue of extreme fame leverage, or did this always exist? Should celebrities, instead of promoting their brands, be promoting awareness of issues such as climate change?

84. There are so-called classic films like *Aladdin* and *Swiss Family Robinson* on Disney's streaming service. Because some of these Disney films portray harmful stereotypes, before viewers watch some of these films, they are warned with a disclaimer, which states: "These stereotypes were wrong then and are wrong now. Rather than remove this content, we want to acknowledge its harmful impact, learn from it and spark conversation to create a more inclusive future together."

Discuss your thoughts on this disclaimer.

What is a stereotype? If you were a Disney executive, would you remove the films from circulation? Should these films which portray harmful stereotypes be played in schools? What are other ways Disney could mitigate the harmful effects of such portrayals of marginalized groups? Do you agree with how the disclaimer is phrased?

Source: https://www.nytimes.com/2020/10/18/business/media/disney-plus-disclaimers.html

85. Parents who share pictures and videos of their children online (e.g., social media) are sometimes accused of oversharing or "sharenting", that is, using cute or embarrassing moments to boost views, likes and sometimes even income. But there are growing concerns about the impact on children.

What are the potential benefits and harms of this practice?

Can children consent to participating in social media? Should laws be enacted to protect the children of social media influencers? In addition to laws, how else could these impacts be mitigated?

86. Facebook is a social networking platform founded by Mark Zuckerberg in 2004. With over 2.8 billion monthly active users, it has become one of the largest social media platforms in the world, allowing users to connect with friends, family, and communities, as well as access news and entertainment content. However, the platform has also faced criticism.

What ethical considerations should companies such as Facebook take into account when making decisions about data collection and data sharing?

Should Facebook be allowed to share its user data? How should Facebook customers be informed about data collection and sharing policies? What are potential implications if data collection and sharing policies are breached? Who owns information posted on Facebook?

---

87. *NY Med* was a documentary TV series that aired on ABC from 2012 to 2014. It followed the lives of doctors, nurses, and patients at hospitals in New York City. The show offered a unique and often dramatic look into the world of healthcare, showcasing the highs and lows of working in a busy hospital environment. Anita Chanko's husband, Mark, was hit by a sanitation truck in Manhattan. His death was later shown on a TV episode of *NY Med* without his family's permission. Although her husband was blurred out, according to *The New York Times*, "there was no doubt in her mind: the blurred-out man moaning in pain was her husband of almost 46 years, the Korean War veteran she met in a support group for parents without partners."

Discuss the ethical issues in this case.

Should consent have been obtained from Anita? Should real-life medical series be allowed on television? What safeguards could be put in place to ensure scenarios like Anita's don't happen again? Which stakeholders should be involved in resolving these scenarios? Do these medical drama shows have any benefits?

Source: https://www.nytimes.com/2015/01/04/nyregion/dying-in-the-er-and-on-t
v-without-his-familys-consent.html

88. In the United States, the Automotive Coalition for Traffic Safety and the U.S. National Highway Traffic Safety Administration are working together to explore technologies that might be used to combat drunk driving. The two groups are part of the Driver Alcohol Detection System for Safety (DADSS), a research program that is proposing two unique technologies. One technology is a breathalyzer unit fixed to the steering wheel that can test blood alcohol levels of the driver as they exhale normally. The other is a skin-based test that requires the driver to place their finger on a sensor.

Discuss the potential implications of these technologies.

Do these technologies have the potential to be abused? What if the technologies have false positives or false negatives? What are other ways to combat drunk driving? Which stakeholders should be involved in combatting drunk driving?

89. C. Henegan once said: "Life was so simple when apples and blackberries were fruit, a tweet was the sound of nature, and facebooks were photo albums."

Discuss the pros and cons of modern technology.

Provide an example of a technology that was designed to make something simpler, but in fact, made things more complicated. Does every generation think that life was simpler "in the good old days"? What do you think is the most impactful technological advance of the last 10 years? What do you think is the most impactful technological advance of the last 10 years in healthcare?

---

90. In the 1960's, NASA scientists realized that pens could not function in space. They needed an alternate solution for their astronauts to write in space. They spent millions of dollars developing a pen that could write without gravity. The Soviets, however, came up with a simpler solution—and used pencils.

Name a time in your life when you found that a simpler solution was better than a complex one.

Name another time in your life when a more complex solution was better than a simple one. "Occam's razor" is a maxim that can be interpreted as "the simplest of two or more competing theories is preferable"; how can Occam's razor be used to evaluate competing hypotheses in scientific research? Can Occam's razor be applied outside the realm of science, to everyday decision making? Are there any limitations to Occam's razor?

---

91. Filters on social media have become ubiquitous, transforming the way influencers present themselves online. With a tap, influencers can change their appearance, add whimsical effects, or transport themselves to exotic locations.

Is it ethically justifiable for social media platforms to use filters?

Should influencers be required to disclose when they have used filters in their posts? What responsibility do social media platforms have in terms of regulating the use of filters to mitigate the potential harms they cause? Should there be age restrictions in place for the use of certain filters on social media, especially those that significantly alter a person's appearance?

Chapter 8: General Topics

92. NASA (National Aeronautics and Space Administration) is a United States government agency responsible for the nation's space program and aerospace research. It was established in 1958 and has since led many groundbreaking missions, including sending humans to the Moon and Mars.

Is space exploration a waste of money?

What are the benefits of space exploration? How can the value of space exploration be balanced against other societal priorities, such as addressing poverty and inequality, combating climate change, and ensuring access to healthcare and education? Is any field of science that doesn't improve human health worth exploring? What impact has space exploration had on global cooperation and diplomacy, and how has it facilitated international collaboration on important scientific projects?

93. Drug Abuse Resistance Education (DARE) is a school-based program that aims to prevent drug abuse among young people. The program was developed in the United States in the 1980s and has since been adopted by schools in many countries worldwide. DARE is typically delivered by trained police officers who visit schools to teach children about the dangers of drug use and how to resist peer pressure to experiment with drugs. The program includes role-playing activities, group discussions, and interactive exercises designed to increase students' knowledge

and skills related to drug resistance. However, the effectiveness of DARE has been debated, and some studies have found little to no impact on reducing drug use among students.

Why do you think the DARE program hasn't been effective?

If you were a high school principal, would you support the implementation of DARE in your high school? Besides police officers, who do you think could be better suited to teach the program? Besides a school setting, what other setting could be better suited to implement the program? At what age do you think programs like DARE should be implemented?

---

94. The involvement of police officers in responding to mental health crises is a complex issue. While police officers are often the first to respond to 911 calls related to mental health emergencies, they may not have the specialized training or resources needed to effectively de-escalate the situation. In some cases, police officers may resort to using force or arrest as a means of handling the situation.

Is force or arrest an appropriate way to handle mental health crises?

Who could be involved to work with police to be better equipped to handle these situations? What other strategies are needed to deal with mental health issues? Did the COVID-19 pandemic exacerbate mental health crises?

---

95. Elementary school libraries play a crucial role in promoting literacy and a love of reading among young children. They provide students with access to a wide range of books, magazines, and other reading materials that can help to develop their language skills, vocabulary, and imagination. By encouraging children to read

for pleasure, elementary school libraries can also foster a lifelong habit of reading, which has been shown to have a positive impact on academic success and overall well-being. Additionally, elementary school libraries can serve as a valuable resource for teachers, providing them with access to materials that can enhance their lessons and support student learning.

What considerations should be taken into account when determining which books should be included or excluded from a public school library?

Should parents be allowed to decide which books are included? Should the same considerations be taken into account when deciding what internet content children should be able to access?

---

96. Affordable daycare is an important issue for families, especially for those with young children or infants. Access to affordable daycare can help parents to continue working or pursue education while ensuring their children are receiving high-quality care.

What responsibilities do governments have in ensuring that affordable daycare is accessible to all families?

How can these responsibilities be balanced with concerns about cost, quality, and availability of childcare workers? What responsibilities do employers have in ensuring affordable daycare? How do you think the high costs of daycare affect diversity in the workplace?

---

97. Age cut-offs for driving are often a topic of debate in discussions on road safety. While many older adults are safe and competent drivers, age-related changes in

vision, cognitive function, and physical ability can increase the risk of accidents. Age cut-offs for driving are intended to protect public safety by ensuring that drivers have the necessary skills and abilities to operate a vehicle safely. However, it's important to recognize that age cut-offs can also have unintended consequences.

What are some potential unintended consequences of age cut-offs for driving?

There is currently no age cut-off preventing older adults from driving in many countries; should there be a mandatory cut-off? To obtain a driver's license, there is usually a minimum age (e.g., 16 years old); does a minimum age make sense for <u>obtaining</u> a driver's license? What criteria could be used instead of age in determining whether one should be allowed to hold a driver's license?

---

98. Open data refers to the practice of making data freely available to the public, allowing anyone to access, use, and share the information. The primary aim of open data is to increase transparency and accountability in government and public organizations, promote innovation, and facilitate evidence-based decision making. Open data covers a wide range of topics and can be used for various purposes, from improving public services and infrastructure to advancing research and development. Many governments, international organizations, and private companies have embraced open data initiatives, leading to the creation of numerous open data portals and platforms worldwide.

What benefits could arise from open data?

What are potential concerns of open data? Do you think this should apply to health and research data? Does it make a difference if data are de-identified? What do you think is "de-identified" data?

99. The Sikh faith requires its adherents to wear a turban as a symbol of their faith and identity. However, wearing a turban can make it difficult to fit and wear a motorcycle helmet, which is required by law in many countries. In response, some Sikh riders have sought exemptions from helmet laws on religious grounds.

Should judges allow exemptions from these laws?

What are the alternatives? What other examples exist where safety intersects with religious freedom? How could this situation lead to innovation in helmet design?

100. Kirpans are ceremonial daggers that are a central symbol of the Sikh faith. They are typically worn by baptized Sikhs as an article of faith and represent a commitment to upholding justice, defending the weak, and living a life of courage and sacrifice. However, the wearing of kirpans has sparked controversy in some educational institutions since school policies prohibit students from carrying weapons.

Should schools make exemptions to the policy?

How can religious freedoms of individuals or groups be balanced against the ethical responsibility to maintain safety and protect the well-being of society as a whole? Should religious institutions be held accountable for violating safety regulations, even if doing so is seen as necessary to practice their faith? What other examples exist where safety intersects with religious freedom?

---

101. Some have argued that actors who play disabled characters in film or TV shows should themselves be disabled.

Discuss your thoughts on this statement.

What are the benefits of including disabled actors? How does this relate to equity, diversity, and inclusion? Do similar arguments hold for transgender roles and transgender actors? Do similar arguments hold for racialized roles and racialized actors?

---

102. A "fat tax" is a proposed tax on foods and beverages that are high in calories, sugar, and fat, with the aim of discouraging people from consuming unhealthy products and reducing rates of obesity. There is debate over the effectiveness of a fat tax, with some studies suggesting that such taxes may not have a significant impact on consumer behaviour or health outcomes.

What are the potential implications of this type of tax?

How could the benefits of a fat tax be measured? What are examples of similar taxes? Have these other taxes been effective?

---

103. "Food security" is said to exist when people have reliable access to nutritious food for a healthy life.

Do you think there are other important components of food security in addition to access and nutrition?

How can we tell when an individual or community is "food secure"? Why is food security important? Do you think most people in your country are food secure? How do you think food security relates to poverty?

---

104. "Voluntourism" refers to the practice of combining travel with volunteer work, usually in developing countries. Some examples include volunteering in orphanages or volunteering for short-term building projects. Proponents of voluntourism argue that it provides a unique opportunity for individuals to make a positive impact and contribute to local communities. However, critics argue that voluntourism can be problematic.

What are potential benefits of voluntourism?

What are potential harms of voluntourism? Should healthcare professional schools require a form of voluntourism prior to entry to their school? Should governments regulate voluntourism?

---

105. Hormone testing in Olympic athletes is a controversial issue, as it involves the use of medical procedures to determine whether athletes have natural hormone levels that fall within a certain range. The International Olympic Committee (IOC) has implemented regulations that limit the levels of testosterone in female athletes in order to prevent those with naturally high levels from gaining an unfair advantage over their competitors.

Do you agree with the IOC regulations?

Are there any potential dangers of hormone testing? What are alternatives to hormone testing? Should the same regulations apply to amateur sports? Should all sports be treated equally when it comes to hormone levels (e.g., weightlifting versus golf)?

---

106. Using cell phones while driving has been identified as a significant cause of traffic accidents and fatalities. Studies have shown that drivers who use cell phones are more likely to be involved in accidents due to distraction. In response to this, many countries have implemented laws banning or limiting the use of cell phones while driving.

Do you agree with laws to ban cell phones while driving?

What are the challenges of enforcing these laws? Who should be exempt from these laws? What are alternatives to laws to deter the use of cell phones while driving?

---

107. Restorative justice is a theory and practice of justice that focuses on the harm caused by a crime or conflict, and seeks to repair the harm through dialogue, understanding, and mutual agreement. Unlike traditional punitive justice, which focuses on punishment and retribution, restorative justice aims to foster accountability, healing, and restoration for all parties involved, including victims, offenders, and the community. Restorative justice practices can take various forms, including victim-offender mediation, community conferences, and circles. The approach has gained recognition and support in many countries, as it offers an alternative to the punitive justice system.

Do you think restorative justice should replace the prison system?

Instead of replacing the prison system, could restorative justice complement the prison system? Do you think restorative justice would be effective for all individuals? How can restorative justice be best implemented? Do you think there are any potential negative consequences of some restorative justice practices?

---

108. Homework in elementary school education is a topic of ongoing debate among educators and parents. Some argue that homework has benefits, while others argue that homework is not appropriate for elementary school-aged children.

What is the purpose of homework?

Should teachers limit the amount of homework? How could socioeconomic status affect the effectiveness of homework? In what grade should homework be started? Should homework be prioritized over physical activity? Should quality or quantity be emphasized for homework?

---

109. *The World Happiness Report* has established a ranking system for the happiness levels of hundreds of countries.

How do you think happiness is defined in this report?

What are challenges in measuring happiness? What is your definition of happiness? Do you think happiness changes over time? What do you think determines happiness? Are you more willing to visit a country with a high happiness rank?

---

110. With the recent death of Queen Elizabeth II, the Bank of Canada will make a decision on who will replace the monarch on the $20 bill. The Bank of Canada has a shortlist of eight "bankNOTE-able Canadians" who are contenders to be the new face of the bill, narrowed down from more than 600 people nominated by the general public.

Who would you nominate to be on the new Canadian $20 bill (or a new bill in your country)?

Why would you nominate this person? If you could only nominate a scientist or healthcare professional, who would you nominate to be on the new bill? If you could only nominate someone in your peer circle or family, who would you nominate to be on the new bill? If individuals were not eligible to be on the bills, what symbol or animal would you choose?

---

111. In his book *The Organized Mind*, neuroscientist Daniel Levitin states that the average supermarket in 2014 stocked 40,000 items, compared with just 9,000 items in 1976. These extra options mean that the modern consumer must make more decisions than ever before.

How do you make a decision when faced with multiple options?

Do you think more decisions leads to more stress? Do you think we are better off with more decisions in a supermarket? Do you think we are better off with more decisions in terms of treatment options for a disease?

---

112. George Orwell wrote extensively on power and its role in society, particularly in his famous novel *1984*. According to Orwell, power is not just the ability to control

others through force or coercion, but also the ability to control language and information. In *1984*, the government's power is maintained through the manipulation of language and the erasure of history, which allows the government to control people's thoughts and beliefs.

What do you think is the relationship between language and power?

How can language affect power status? How can power status affect language? What is the relationship between language and thought? Why is erasure of history problematic?

113. Greta Thunberg is a Swedish environmental activist who gained international attention for her climate change activism, including leading the global "Fridays for Future" school strike movement. In 2019, Thunberg was nominated for the Nobel Peace Prize for her activism. Although she did not win the prize, her nomination brought global attention to the urgent need for action on climate change. Thunberg continues to be a leading voice in the climate movement, and her advocacy has inspired millions of young people around the world to take action on environmental issues.

What is the relationship between climate change and peace?

What are the qualities of a good youth advocate? What role should healthcare providers play in advocating for climate change? What are the qualities of an advocate within the context of healthcare?

114. There is no question that there have been lasting changes from the COVID-19 pandemic.

What was the most long-lasting change of the pandemic?

How did the pandemic affect employment? How did the pandemic affect health and healthcare? How did the pandemic affect travel? How did the pandemic affect education? Of all these sectors, which was most affected by the pandemic?

---

115. Peanut allergies are a common and potentially life-threatening food allergy. Symptoms of a peanut allergy can range from mild, such as hives or a rash, to severe, including anaphylaxis, a potentially fatal reaction. Peanut allergies are on the rise, and there is currently no cure. The best way to prevent a peanut allergy reaction is to avoid peanuts and peanut products, and some schools ban peanut products in the classroom. In some countries such as Canada, parents have filed human rights complaints, claiming that elementary schools don't do enough to protect against allergic reactions to foods such as peanuts.

Discuss your thoughts on these complaints.

How far should schools be required to go to accommodate food allergies? Should other allergies such as to gluten or soy also be accommodated in schools? Should workplaces also be obliged to accommodate food allergies?

---

116. With increasing competition for spots in university, many parents decide to pay their children for good grades in high school. Offering financial incentives for good grades can have both pros and cons.

What are some of the pros and cons of paying kids for good grades?

Instead of financial incentives, what other incentives could parents use to motivate children? What incentives could teachers use to motivate children? Do you think children who are paid for good grades will be successful later in life?

---

117. Social isolation is a public health concern and has been associated with increased risk of cardiovascular disease among older adults.

Why do you think social isolation affects cardiovascular health?

What are potential solutions to social isolation? How could mobility affect social isolation? How could hearing and vision impairments affect social isolation? Do you think social isolation could lead to dementia? What are other potential causes and consequences of social isolation? How did the COVID-19 pandemic compound the issue of social isolation?

---

118. During the COVID pandemic, several hospitals offered free parking.

Should hospitals offer free parking?

To whom should free parking be offered? To whom should the profits of parking be given? Should staff or patients be given priority to access parking? Should individuals who can't afford hospital parking be compensated?

---

119. Tobacco use is the leading cause of preventable death worldwide, and smoking is a major risk factor for lung cancer. When tobacco is smoked, it releases harmful chemicals that can damage lung tissue and DNA, leading to the development of

cancerous cells. The longer and more frequently a person smokes, the greater their risk of developing lung cancer. While quitting smoking can reduce the risk of lung cancer, it can take many years for the risk to return to that of a non-smoker.

Should the tobacco industry compensate people with lung/throat cancer and emphysema?

How do we know when a factor (like smoking) causes a health outcome like cancer? What are challenges in deciding when a factor (like smoking) causes a health outcome? What role do healthcare providers have in preventing or treating tobacco-related illnesses?

---

120. According to the Centers for Disease Control and Prevention (CDC)

> [c]limate change, together with other natural and human-made health stressors, influences human health and disease in numerous ways. Some existing health threats will intensify and new health threats will emerge. Not everyone is equally at risk. Important considerations include age, economic resources, and location.

What are some of the ways that climate change could influence health?

Why do age, economic resources, and location change the risk of health threats? Are there other factors that might also affect the risk of health threats? Should healthcare professionals advocate to protect the environment?

Source: https://www.cdc.gov/climateandhealth/effects/default.htm

121. Leadership is the process of guiding and directing others towards a common goal. It involves the ability to inspire, motivate, and influence people to achieve their full potential.

What are the qualities of a leader?

Describe a time in your life when you exemplified these qualities of a leader. Describe a time in your life when you were inspired by a good leader. What is the difference between a coach and a leader?

---

122. In warm countries around the world, clothes are often dried outdoors under the sun. However, in several areas, bylaws prohibit residents from setting up clotheslines or drying racks outside. The City Council has decided to host a town hall to hear views from community members on this issue.

What argument would you present to City Council in favour of outdoor clotheslines?

What argument would you present to City Council in opposition of outdoor clotheslines? How would you convince skeptical Council members to agree with you? How would you react if all Council members disagreed with you?

---

123. Resiliency training is the process of developing mental and emotional toughness to cope with and recover from adversity. It involves learning skills and strategies that enhance personal resilience and the ability to bounce back from challenging situations.

What are the potential benefits of resiliency training?

Which professions could benefit from resiliency training? What types of skills do you think should be part of resiliency training? Do you think resilience can be taught? What are the challenges in measuring the success of resiliency training?

---

124. Communication was severely affected by the COVID-19 pandemic. Protective measures, such as face masks and physical distancing, were essential to control the spread of the virus, but posed challenges to face-to-face communication.

What are some of the challenges of wearing masks in face-to-face communication?

What are some of the challenges of wearing masks in face-to-face communication for young children, particularly those learning language skills? What are some of the challenges of wearing masks in face-to-face communication in a healthcare setting? Are there cultural differences in non-verbal communication?

---

125. Dress codes in schools are a set of guidelines that regulate what students are allowed to wear on campus or school grounds. They can vary greatly from school to school and may include restrictions on clothing items such as shorts, tank tops, and hats.

Do you agree with having school dress codes?

What is the purpose of a dress code? Who should enforce school dress codes? How could dress codes be harmful and how could they be beneficial? How could dress codes interfere with individual freedoms?

---

126. "Cancel culture" refers to the phenomenon of publicly denouncing and boycotting individuals or entities for controversial statements or actions. It involves swift and widespread condemnation, often through online social media channels, leading to potentially severe consequences.

What are some of the severe consequences of cancel culture?

Is cancel culture an effective means of holding individuals accountable for their actions, or does it encourage a culture of online vigilantism? Is it fair for cancel culture to target individuals who have apologized for their actions? Does cancel culture allow for the possibility of personal growth, redemption, and forgiveness? What impact does cancel culture have on freedom of speech? Can you think of any modern examples of cancel culture? Did social media create cancel culture? Did cancel culture exist before social media?

---

127. "Armed conflict" refers to a state of violent confrontation between organized groups or nations involving the use of weapons and military force. Armed conflicts have wreaked havoc, leaving behind untold suffering, displacement, and loss of life. Researchers have recently documented that armed conflict has been in decline over the past few decades. Researchers have also attempted to predict countries at higher risk of armed conflicts by looking at an array of country-level data.

What do you think are factors that predict which countries are at higher risk of future armed conflict?

Why do you think armed conflicts have declined over the past few decades? What do you think is the role of healthcare providers in armed conflicts?

128. Obstacles can hinder or obstruct progress in our lives.

Describe a time in your life when you overcame a major obstacle.

What tools did you use to overcome the obstacle? Did you ask for help to overcome the obstacle? How do you think what you learned from overcoming an obstacle could be applied to your journey as a healthcare professional? Do you think overcoming an obstacle makes you more resilient?

---

129. Imagine you could sit down and have dinner with anyone from history.

Who would it be and why?

What questions would you ask? What would you prepare for them for dinner? Is there anyone else you would invite to the dinner?

---

130. Success can be defined in many ways.

How do you define success?

How do you think you will define success in a healthcare professional school? Do you think your definition of success has changed over the years? How does success relate to happiness?

---

131. Goal setting is the process of defining and establishing specific, measurable, achievable, and time-bound objectives. It helps individuals and organizations

identify what they want to achieve and determine the necessary steps to reach those objectives. Effective goal setting provides focus, motivation, and a roadmap for success.

Describe a time you set a challenging goal for yourself.

Did you accomplish the goal? Did you work alone when trying to accomplish the goal or did you work as a team? Why do you think it is important to have measurable objectives? Why do you think it is important to have specific objectives?

Chapter 9: Interpretation & Teaching Topics

132. In Rudyard Kipling's famous poem "The Ballad of East and West," he writes: "Oh, East is East, and West is West, and never the twain shall meet."

Describe a situation in your personal or professional experience where unity or agreement seemed impossible.

How did you resolve the situation? Did you consult anyone to get help in resolving the situation? What did you learn from the situation that you were able to apply in another situation? How can individuals and groups work to build bridges and find common ground even when they hold vastly different beliefs or perspectives? What role can active listening, open communication, and empathy play in promoting unity in the face of disagreements? In what ways can disagreements actually strengthen relationships and lead to more effective problem solving and decision making?

133. In a famous poem, Robert Frost wrote:

> I shall be telling this with a sigh
> Somewhere ages and ages hence:
> Two roads diverged in a wood, and I—

I took the one less traveled by,
And that has made all the difference.

Describe a time in your life when you took the road "less travelled."

What does it mean to take the road less travelled, and what are some potential benefits and drawbacks of doing so? How can taking the road less travelled lead to personal growth and development, and in what ways can it challenge individuals to step outside of their comfort zones? In what ways can taking the road less travelled contribute to creativity, innovation, and original thinking? What roles can risk-taking and courage play in taking the road less travelled, and how can individuals overcome fear and uncertainty in order to pursue unconventional paths? What are some real-world examples of individuals or groups who have taken the road less travelled, and what can we learn from their experiences?

---

134. "Before you judge someone, walk a mile in their shoes."

Discuss how you think this quote applies to training and practising in healthcare professions.

What is empathy, and how does it differ from sympathy or pity? How can empathy be cultivated and developed, and what benefits can it offer to individuals and society as a whole? In what ways can empathy help us to better understand and connect with others, particularly those who come from different backgrounds or have different experiences? What are some potential challenges or drawbacks of empathy, and how can these be addressed or mitigated? How might empathy play a role in promoting social justice and addressing systemic inequalities such as racism or economic inequality?

135. In Rudyard Kipling's poem, he writes:

> If you can meet with Triumph and Disaster
> And treat those two impostors just the same

How does the speaker advise the reader to treat triumph and disaster?

According to Kipling, why are triumph and disaster both considered "imposters"? How might one's ability to treat triumph and disaster equally contribute to personal growth and character development? In what ways might the idea of treating triumph and disaster as impostors be applied to real-life situations and challenges?

---

136. You have been asked to teach about insects to a class of Grade 1 students, ages 6 to 7.

What would you include in your lesson plan?

How will you prepare your lesson plan? How will you modify your language for this age level? Would you include any visuals or hands-on activities?

---

137. Using only the pencil and paper provided, draw three things to represent your desired healthcare specialty (e.g., midwifery, medicine, nursing, dentistry, pharmacy, or physiotherapy). Don't worry about your drawing skills!

Explain your drawing to the interviewer.

How does the concept of teamwork fit into your drawing? How does the concept of equity, diversity, and inclusion fit into your drawing?

About the Author

As a faculty member at the University of British Columbia in Canada, Dr. Jacquelyn Cragg, PhD has educated and mentored a range of healthcare professionals and pre-healthcare students, from pharmacy to medicine to nursing. She is the recipient of numerous awards, including a Tier 2 Canada Research Chair, Michael Smith Health Research BC Scholar Award, and L'oréal UNESCO for Women in Science International Rising Talents prize.